ULTIMATE SUPERCARS

BMW i8

By Tamra B. Orr

Kaleidoscope
Minneapolis, MN

The Quest for Discovery Never Ends

....

This edition first published in 2023 by Kaleidoscope Publishing, Inc.

No part of this publication may be reproduced in whole or in part without written permission of the publisher.

For information regarding permission, write to Kaleidoscope Publishing, Inc.
6012 Blue Circle Drive
Minnetonka, MN 55343

Library of Congress Control Number
2022937981

ISBN
978-1-64519-607-5 (library bound)
978-1-64519-677-8 (ebook)

Text copyright © 2023 by Kaleidoscope Publishing, Inc. All-Star Sports, Bigfoot Books, and associated logos are trademarks and/or registered trademarks of Kaleidoscope Publishing, Inc.

Printed in the United States of America.

Bigfoot lurks within one of the images in this book. It's up to you to find him!

TABLE OF CONTENTS

Chapter 1: The Force of Tomorrow **4**

Chapter 2: From Aircraft to Cars **12**

Chapter 3: Unpacking a Hybrid **18**

Chapter 4: The First of Its Kind **24**

Beyond the Book *28*
Research Ninja *29*
Further Resources *30*
Glossary *31*
Index *32*
Photo Credits *32*
About the Author *32*

Chapter 1
The Force of Tomorrow

The actor looks straight into the camera in one of BMW i8's first commercials. He looks serious. He wants your attention. Then he talks to you as if he is speaking for the unusual car.

"I am changing the game for all you warriors, doubters ..." he states. "I am unstoppable, a rocket, a cannonball, a **carbon-fiber** body lighter than wind, stronger than storm ... I am born electric, the force of tomorrow."

A smooth, shiny BMW i8 races down the road. The actor adds, "I am possible."

The message was clear. BMW had quite the car!

FUN FACT
The quotes are from an actor on one of the first i8 commercials made by BMW.

BMW has been making cars for almost one hundred years. The i8 came out in mid-2014. It was a **hybrid**, unlike the company's other cars. Its gas-powered engine was paired with an electric motor. Hiding inside the car were the two engines as it raced down the road with the power usually found in race cars. Super-fast, high-tech, and still good for the **environment**? That is a rare combination. Yet, BMW did it. As the commercial said, they changed "impossible" to "possible"!

PARTS OF A BMW i8

Gorilla Glass bulkhead

kidney grille

low-slung hood

laser headlights

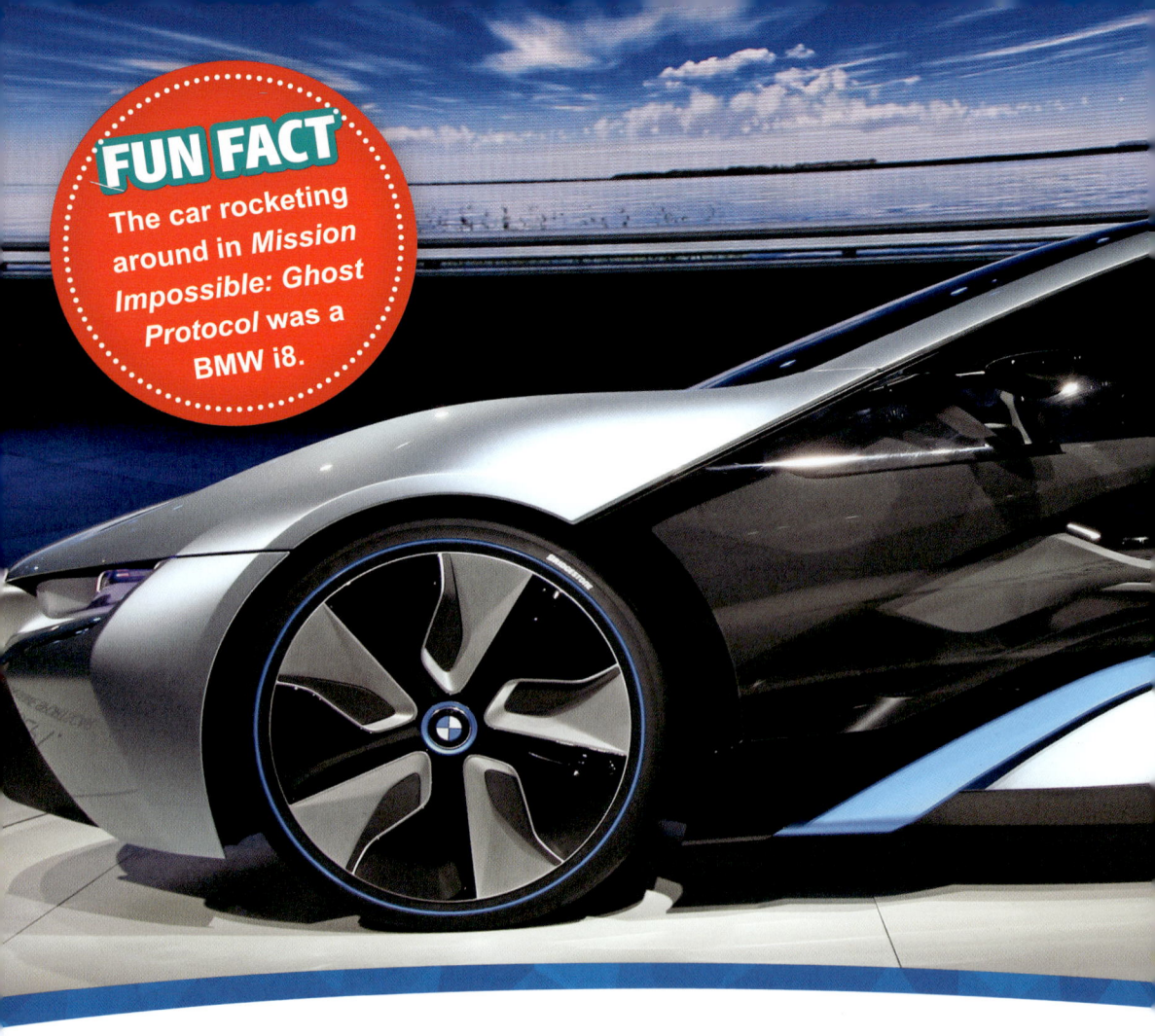

FUN FACT

The car rocketing around in *Mission Impossible: Ghost Protocol* was a BMW i8.

It's clear that the i8 is a close cousin to a race car. It is not as fast, but it is extra light and has a low, wide front base. This makes it more **aerodynamic**. By reducing the push of air flowing past the car, it can go further and faster. The i8 has a narrow kidney-shaped grille. Since the gas engine is not in the front, it does not need air to cool it.

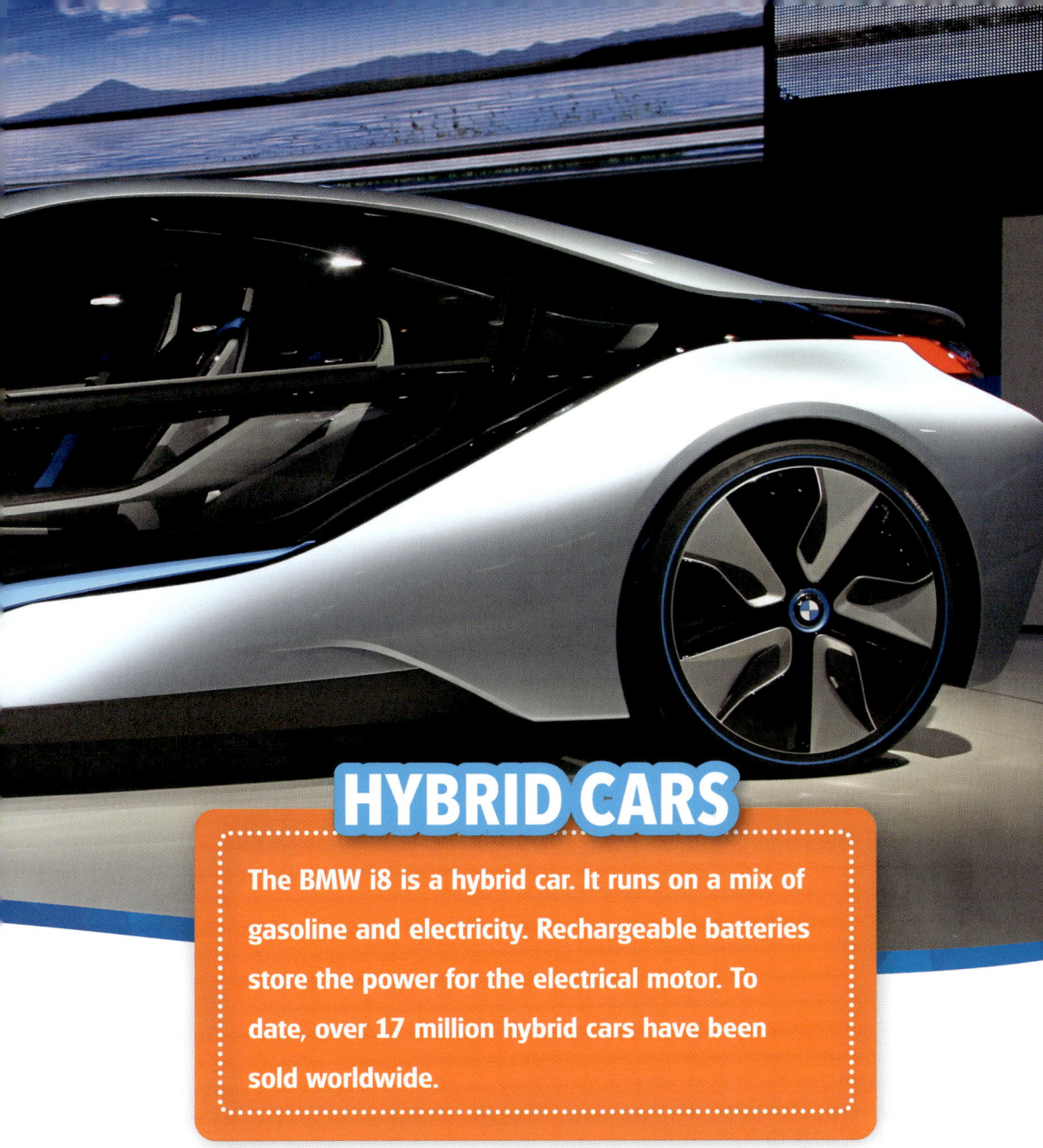

HYBRID CARS

The BMW i8 is a hybrid car. It runs on a mix of gasoline and electricity. Rechargeable batteries store the power for the electrical motor. To date, over 17 million hybrid cars have been sold worldwide.

The car's weight is divided in half between the **axles**. It is easy to handle on curves, even at high speeds.

Chapter 2
From Aircraft to Cars

All over the world, people recognize BMW vehicles. Millions of cars have been produced. Each year, BMW ships thousands of cars to countries like Japan, Africa, and Australia. Surprisingly, the company started out with aircraft.

In 1913, engineer Karl Rapp worked for a German aircraft company. He saw that the aircraft engines had problems and wanted to fix them. He started his own company. Over the next few years, the company grew. The engines improved. It was not until March 1916 that the name Bavarian Motorworks—or BMW—was formed.

Soon BMW was making engines for aircraft, motorcycles, and cars. That changed when World War I began. Many of BMW's factories were bombed. Production was halted. BMW began using the metal they had to make thousands of pots, pans, train brakes, spare parts, and even bicycles!

By the mid-1920s, BMW was back to making motorcycles and cars. In 1959, a competitor took over the company. Soon more cars were being made, and BMW was making a name for itself.

WHERE ARE BMW'S MADE?

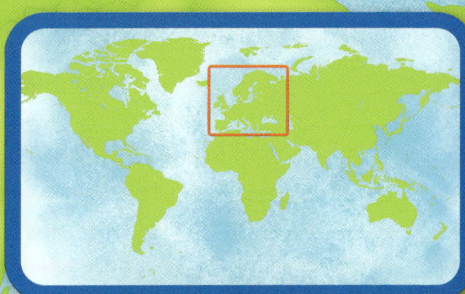

Germany Poland

France

Munich, Germany

BMW TOWER

In 1999, BMW's Headquarters was named a protected historic building. BMW Tower is located in Munich, Germany.

MEET THE 1602 ELEKTRO-ANTRIEB

While the i8 is BMW's most high-tech electric car, it is not the company's first one. In 1972, it produced the 1602 Elektro-Antrieb. Under the hood were a dozen 12-volt batteries weighing over 770 pounds (349.3 kilograms). During Munich's Olympic Games, they used it as a pace car for marathon runners.

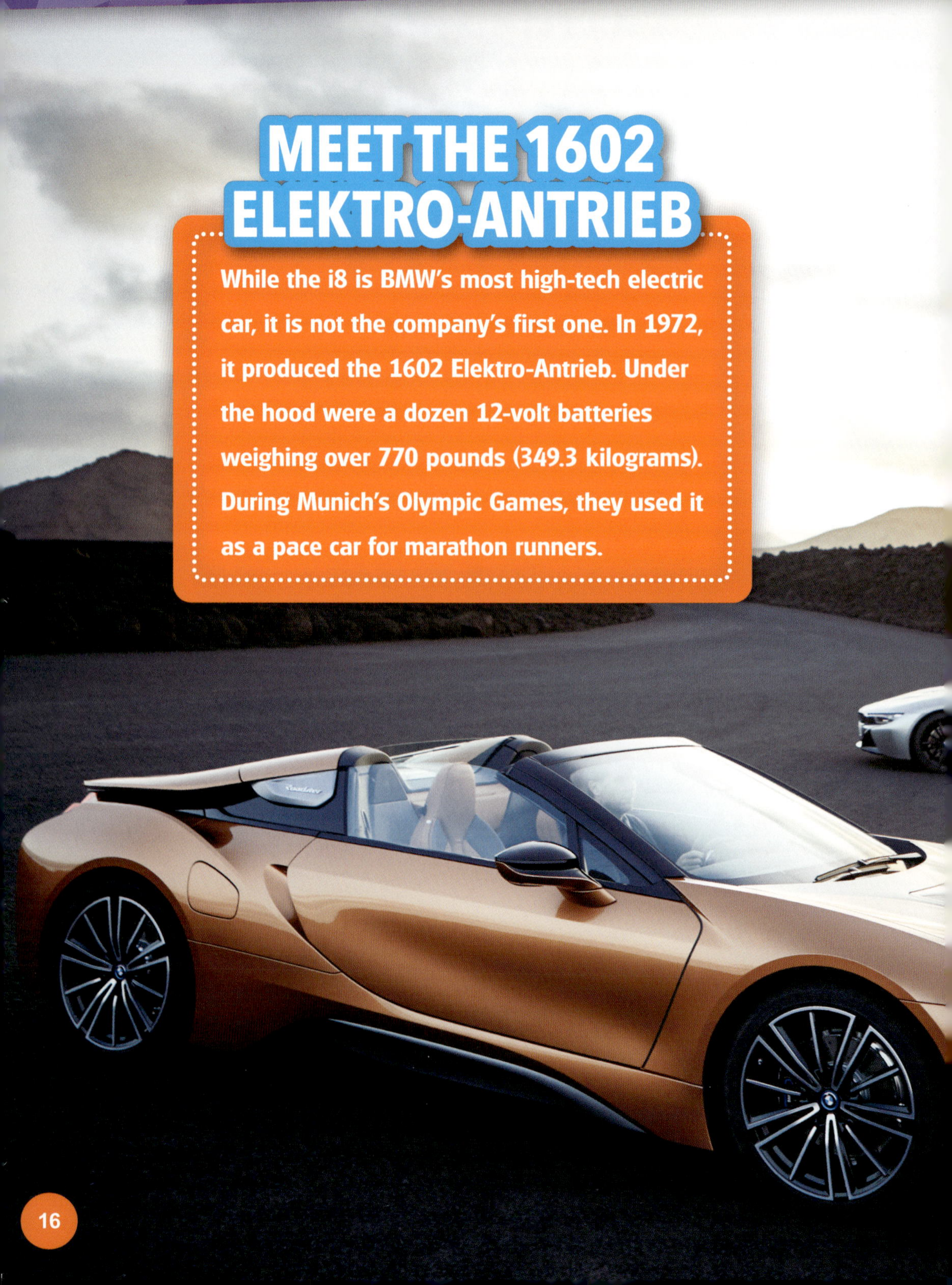

FUN FACT

Some i8 owners miss the traditional roar of a gas-powered engine so BMW made the sound louder to play through the car's speakers.

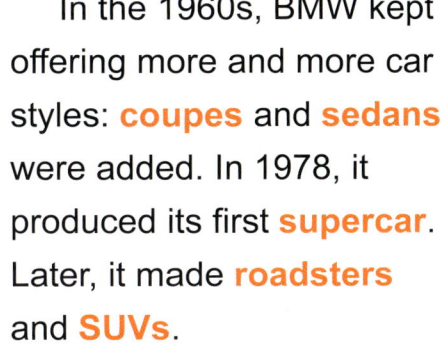

In the 1960s, BMW kept offering more and more car styles: **coupes** and **sedans** were added. In 1978, it produced its first **supercar**. Later, it made **roadsters** and **SUVs**.

In 1994, the European company set up factories in the United States and other countries. Since then, it has grown. Today, it focuses on electric cars.

BMW models have a special logo. The circle has the letters BMW around the top. The blue and white checkers stand for the Bavarian flag.

Chapter 3
Unpacking a Hybrid

BMW released the i8 in 2014. Since then, it has sold more than 20,000 units. It is easy to see why too. It is rare to find a car that mixes beauty and speed with high gas mileage and electricity, but BMW has done it. In fact, it has sold more i8s than all of the other electric sports car companies combined.

One look at the i8 and it almost seems like it came from the near future. Outside it has special scissor doors. Made from aluminum, carbon, and plastic, they weigh half of what normal car doors do.

THE BMW i8 IN DETAIL

COST: $148,495 basic model

Height: 4.2 feet (1.3 m)

Width: 6.4 feet (1.95 m)

LENGTH: 15.43 feet (4.7 m)

WEIGHT: 3,501 pounds (1588.0 kg)

TOP SPEED: 155 miles per hour (249.4 km/h)

TIME FROM 0 to 60 miles per hour: 3.5 seconds

The i8's transmission is unlike most cars. It features front-wheel drive when it is running on electricity. Then it switches to rear-wheel drive when using the gas engine. When both are in use, the i8 becomes all-wheel drive. Laser headlights and taillights use less energy. But they provide more light.

Between where the engine and the passengers sit is a wall made of the same thin, tough glass used in smartphone screens. It reduces noise. Chemicals are used to harden the glass to prevent breaking in an accident.

The i8's trunk is small. The back seat area of the i8 stores the car's fold-up top in the convertible model.

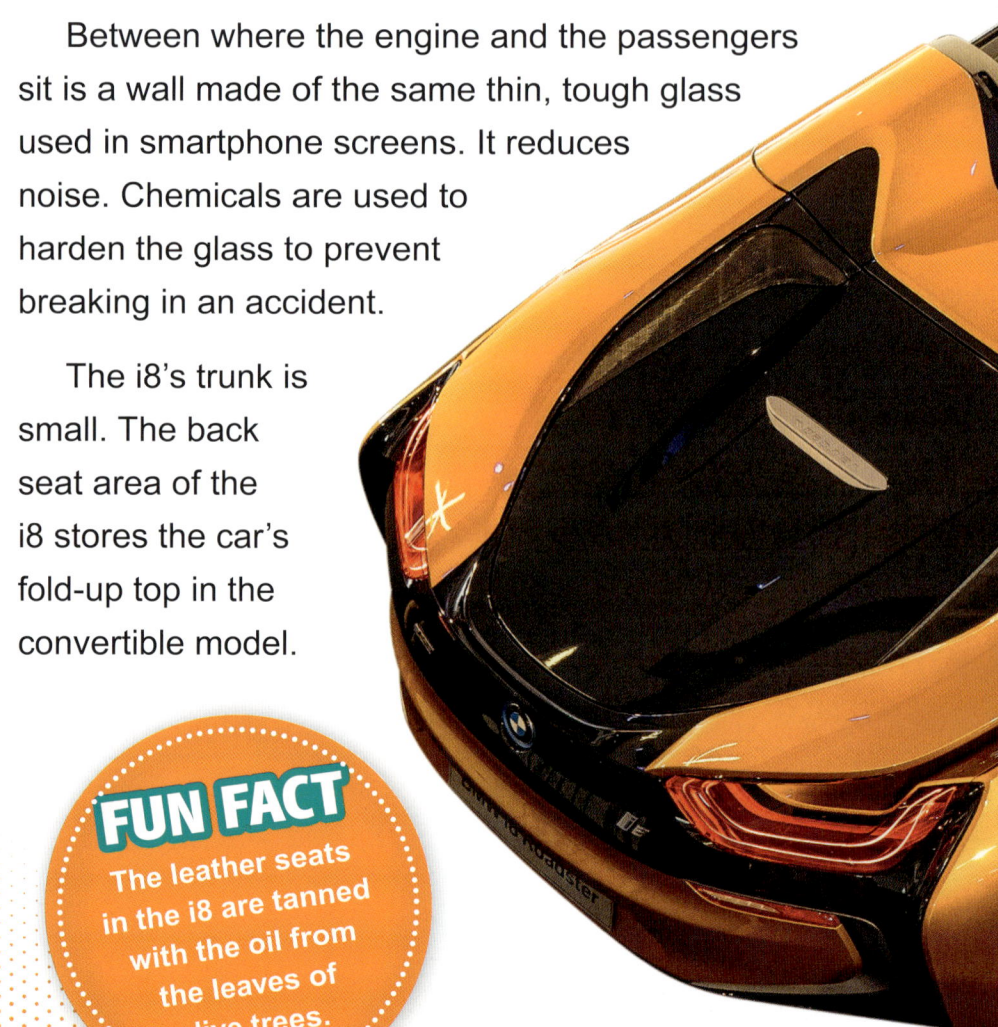

FUN FACT

The leather seats in the i8 are tanned with the oil from the leaves of olive trees.

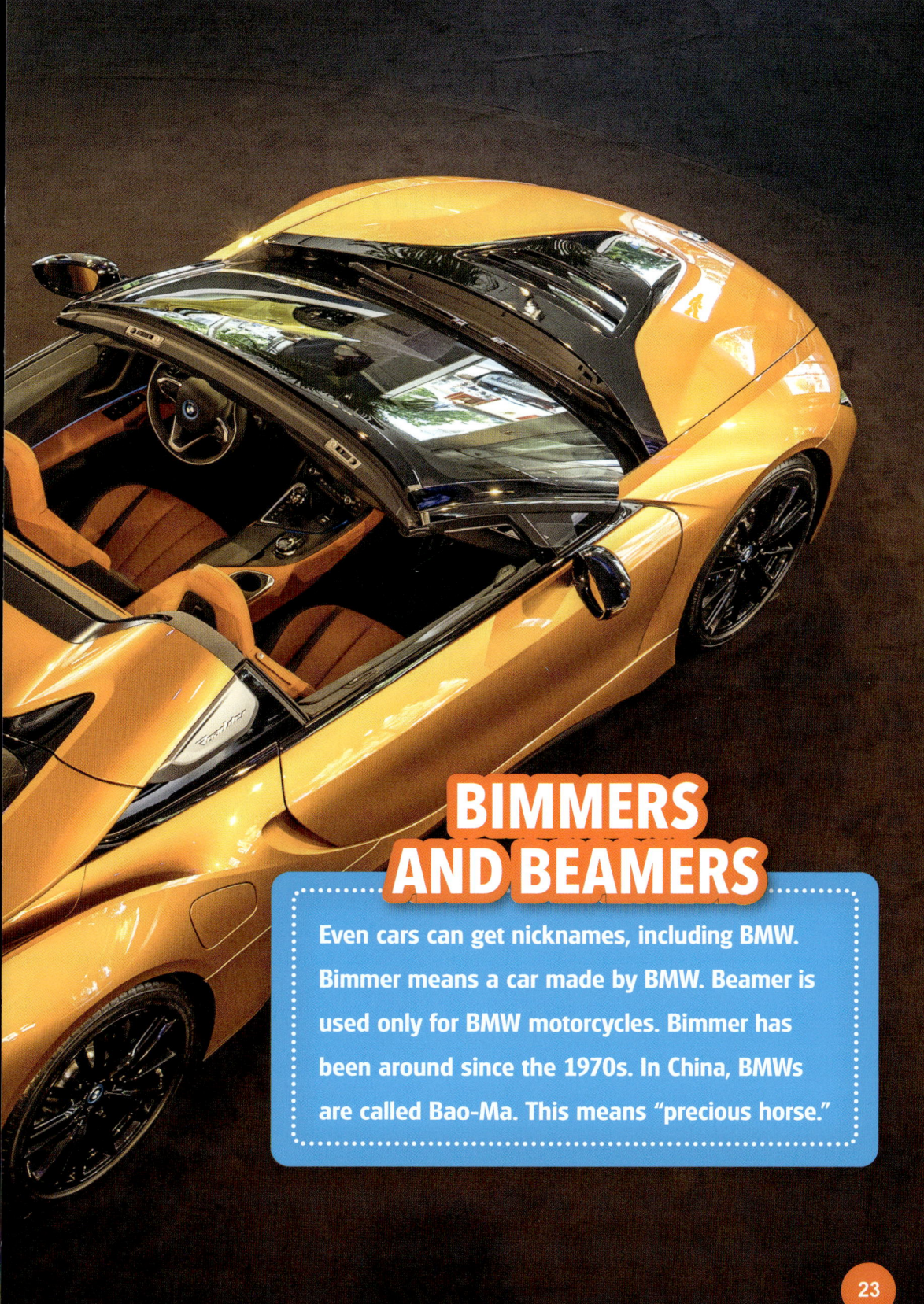

BIMMERS AND BEAMERS

Even cars can get nicknames, including BMW. Bimmer means a car made by BMW. Beamer is used only for BMW motorcycles. Bimmer has been around since the 1970s. In China, BMWs are called Bao-Ma. This means "precious horse."

Chapter 4
The First of Its Kind

The feeling that you are looking into the future keeps going when you slide into an i8. An 8.8 inch (22 cm) media display allows you to listen to your music, plus control the inside temperature or heat up the seats. Check how much charge the battery has. Look at the cameras to park safely. Rely on different alarm systems to warn you if you are getting too close to another car or person.

Although the i8 is an amazing car, its time came to an end in April 2020. BMW decided to focus on making luxury models like the Gran Coupe.

FUN FACT
BMW i8 owners can track their car battery's charge and if the car needs any servicing with their smartphones.

Although people are sad to see the i8 end, they know that BMW will soon follow it up with another great hybrid car. For now, they can see it in the BMW Museum in Munich.

The BMW i8 is one of the most unusual cars in auto history. As Mark Thiesburger from BMW Group Classic stated, "We can already be sure today that this BMW [i8] will become a classic of the future, thanks to many unique properties it has brought together for the first time. In 20 years' time, people will say it was the first of its kind."

THE ULTIMATE SOPHISTO EDITION

Before saying goodbye to the i8, BMW produced 200 of the Ultimate Sophisto model. They matched the gray metallic paint with copper accents and special leather upholstery. Each car came numbered to show it is an extra special edition.

BEYOND THE BOOK

After reading the book, it's time to think about what you learned. Try the following exercises to jump-start your ideas.

RESEARCH

FIND OUT MORE. Where would you go to find out more about your favorite cars? Find out what company makes the car and locate its website. What information do the companies provide? What other sources of car information can you find?

CREATE

GET ARTISTIC. Cars start with creative artists and designers. Time for you to take a shot! Get art materials and create a great, new car. Will you make it a sports car? A sedan? A race car? What colors will you paint it? What features can you give it? Let your imagination go for a spin!

DISCOVER

DIG DEEPER. Hollywood loves supercars. The i8 is in the movie *Mission Impossible: Ghost Protocol*. What other famous movie cars can you think of? For fun, pick your favorite movie and design a car the heroes could use. What features would it have? What would it look like?

GROW

GO TO A CAR SHOW. Car shows are a great way to see lots of cool cars up-close. Check your local events calendar, or ask at a car dealer for upcoming events. You can find shows of old cars and new cars, sports cars and classic cars. Go to a show and find a new favorite car to love!

RESEARCH NINJA

Visit www.ninjaresearcher.com/6075 to learn how to take your research skills and book report writing to the next level!

RESEARCH

DIGITAL LITERACY TOOLS

SEARCH LIKE A PRO
Learn about how to use search engines to find useful websites.

FACT OR FAKE?
Discover how you can tell a trusted website from an untrustworthy resource.

TEXT DETECTIVE
Explore how to zero in on the information you need most.

SHOW YOUR WORK
Research responsibly—learn how to cite sources.

WRITE

GET TO THE POINT
Learn how to express your main ideas.

PLAN OF ATTACK
Learn prewriting exercises and create an outline.

DOWNLOADABLE REPORT FORMS

Further Resources

BOOKS

Colson, Rob. *Custom Cars*. Lebanon, IN: Hachette Children's Group, 2020.

Peterson, Megan Cooley. *BMW i8 (Epic Cars)*. Mankato, MN: Black Rabbit Books, 2021.

Wasef, Basem. *Speed Read Supercar: The History, Technology and Design Behind the World's Most Exciting Cars*. Beverly, MA: Motorbooks, 2018.

WEBSITES

Factsurfer.com gives you a safe, fun way to find more information.

1. Go to www.factsurfer.com.
2. Enter "BMW i8" into the search box and click 🔍
3. Select your book cover to see a list of related websites.

Glossary

aerodynamics: the force experienced by objects moving through air.

axle: a shaft on which a wheel(s) turn.

carbon-fiber: a very strong lightweight synthetic material. The body of a BMW i8 is made of carbon fiber.

coupe: a car with a fixed roof and two doors.

environment: the natural world.

hybrid: a vehicle that runs on electricity and gasoline.

roadster: an open-topped car with a single seat in front.

SUV: a sport utility vehicle used for both daily driving and rough ground.

sedan: a car with a roof, two or four doors, and a separate trunk.

supercar: a street-legal, high-performance sports car.

upholstery: materials used to cover car seats and other types of furniture.

Index

aerodynamic, 10
axles, 11
display, 24
electric, 5, 6, 9, 11, 16, 17, 18
gas, 6, 10, 17, 18, 22
headlights, 8, 22
hybrid, 6, 11, 18, 26
logo, 17

motor, 6, 11
race, 6, 10
Rapp, Karl, 12
supercar, 17
top speed, 20
weight, 11, 20
wheel, 22

PHOTO CREDITS

The images in this book are reproduced through: ezphoto/Shutterstock 3; Fingerhut/Shutterstock 7; North Monaco/Shutterstock 8-9, 10-11, 15; VanderWolf Images/Shutterstock 20; Sergey Kohl/Shutterstock 22-23. All other images courtesy of BMW Group PressClub (Willfried Wulff 6, 14, 16-17, 21).
Cover: Courtesy of BMW Group PressClub, Svetlana Foote/Shutterstock (background).

About the Author

Tamra B. Orr is a full-time author living in the Pacific Northwest with her family. She attended Ball State University before moving cross-country. Orr has written more than 750 books for readers of all ages and says it is the best job in the world because she is always learning something new.